Sunil Lal Rajbhandari
Paras Mani Acharya

Gharial Conservation in Narayani River of Chitwan National Park, Nepal

Sunil Lal Rajbhandari
Paras Mani Acharya

Gharial Conservation in Narayani River of Chitwan National Park, Nepal

LAP LAMBERT Academic Publishing

Imprint

Any brand names and product names mentioned in this book are subject to trademark, brand or patent protection and are trademarks or registered trademarks of their respective holders. The use of brand names, product names, common names, trade names, product descriptions etc. even without a particular marking in this work is in no way to be construed to mean that such names may be regarded as unrestricted in respect of trademark and brand protection legislation and could thus be used by anyone.

Cover image: www.ingimage.com

Publisher:
LAP LAMBERT Academic Publishing
is a trademark of
International Book Market Service Ltd., member of OmniScriptum Publishing Group
17 Meldrum Street, Beau Bassin 71504, Mauritius

ISBN: 978-3-659-63749-0

Copyright © Sunil Lal Rajbhandari, Paras Mani Acharya
Copyright © 2014 International Book Market Service Ltd., member of OmniScriptum Publishing Group

Gharial Conservation in Narayani River of Chitwan National Park, Nepal

Sunil Lal Rajbhandari

Paras Mani Acharya

Acknowledgements

The authors would like to express their sincere thanks to the Rufford Foundation Small Grants Program, U.K. for providing the research grant to carry out gharial research in Chitwan National Park. The Assistant Conservation Officer, Mr. Bed Bahadur Khadka of the Gharial Breeding center, Kasara deserves thanks for his help in coordinating and support during the field activities.

Acronyms

CITES	Convention on International Trade of Endangered Species
CNP	Chitwan National Park
DNPWC	Department of National Parks and Wildlife Conservation
GBC	Gharial Breeding Center
GCA	Gharial Conservation Alliance
IUCN	International Union for Conservation of Nature and Natural Resources
NTNC	Nepal Trust for Nature Conservation
TAL	Terai Arc Landscape
WWF	Worldwide Fund for Nature

Preface

The gharial crocodile is one of the key predators and environmental indicator of the river systems of Nepal which helps in maintaining biotic integrity. The species is listed as critically endangered by IUCN Red Data Book and included as an endangered species by the Nepal Government. The estimated population of gharial crocodile in Nepal is about 80 out of which the Narayani River harbors 44 gharials, the Rapti River, the largest tributary of Narayani supports 24, the Karnali River 8 and the Babai River supports 6 individuals. The population in Narayani is severely threatened by high anthropogenic pressures such as over fishing, extraction of sand and boulders, grazing and dam construction which has led to the fragmentary population as mostly congregated into two groups, namely the larger group in Khoria Muhan and a smaller one in velaunji. Realizing the current declining trend of the gharial population, the government of Nepal through the Department of National Parks and Wildlife Conservation has initiated conservation efforts like the preparation of Gharial Action Plan and regular monitoring of habitats, population and illicit human activities. There has been very little investigation on habitat preference, population, reproductive success and nesting and hatching behavior which have hindered in designing the effective conservation strategies to maintain long term conservation of gharials in the river basins of Nepal.

Table of contents

                                                                            Page No.

Acknowledgements	2
Acronyms	3
Preface-	4
Summary	6
Introduction	7
Rationale of the study	19
Literature review	20
Policies and legislations	30
Objectives	36
Study Area	36
Materials and methods	39
Results	44
Discussion	53
Conclusion and recommendations	55
References	56
Annexes	60-63

Summary

A study carried out in Narayani River of Chitwan National Park from 2012 to 2013 to investigate the habitats and abundances of the gharials has recorded the total number of 38 gharials including 3 hatchlings, 8 juveniles, 12 sub-adults and 15 adults. Only one breeding male was observed reflecting the critical condition for the breeding in the wild within Chitwan Nationl Park. The abundance of the gharial was restricted in 2 areas, namely Khoria Muhan and Velaunji due to suitability of the habitat conditions and low human disturbances. The Basking activities in relation to depth showed that > 80 % of the gharials selected sandy banks. Among the 2 populations, the gharials in Khoria Muhan preferred shallow to moderate depth of water because of presence of braided channels in the area. In the downstream watercourse of Velaunji, the gharials select deep water course of the main channel. The park management must adopt conservation measures such as ban of fishing and regular monitoring of habitats and population, especially protection of nesting sites to maintain long term conservation of gharials in Narayani River.

1.0 Introduction

The Crocodile (Family Crocodylidae) is a large aquatic tetrapod that lives throughout the tropics in Africa, Asia, the Americas and Australia. Lizards, snakes and crocodiles are all scaled diapsids but crocodiles are archosaurs which means they are genetically closer to birds and the extinct dinosaurs. Crocodylidae is classified as a biological family (Bustard and Singh, 1978; Whitaker and Basu, 1983). Adult males grow a bulbous nasal appendage, which resembles a pot called a 'ghara', from which the species derives its name "Gharial".

Historically, *G. gangeticus* was found in the northern part of the Indian subcontinent, in the River systems of Indus (Pakistan), Ganges (India and Nepal), Mahanadi (India) and Brahmaputra (Bangladesh, India and Bhutan). The presence of the species in the Irrawaddy River system in Myanmar has also been reported (Smith, 1931). The Gharial is typically a resident of flowing rivers with deep pools that have high sand banks and good fish stocks (Whitaker and Basu, 1983; GCA, 2008). Exposed sand banks are used for nesting. Although the function of the ghara is not well understood, it is apparently used as a visual sex indicator, as a sound resonator, or for bubbling or other associated sexual behaviors (Martin and Bellairs 1977).

Gharial was previously present in all the Gangetic plain in several countries of South-East Asia – Pakistan, Bhutan, Myanmar (ex-Burma), India, Nepal and Bangladesh. In the 1940's, the gharial population was estimated from 5,000 to 10,000 individuals. It declined to 150-200 individuals in the 1960's principally due to habitat destruction and uncontrolled exploitation. It's now only present in India and Nepal, with about 200 individuals in India and 81 in Nepal (Stevenson & Whitaker, 2010).

Figure 1: Distributional range of gharial

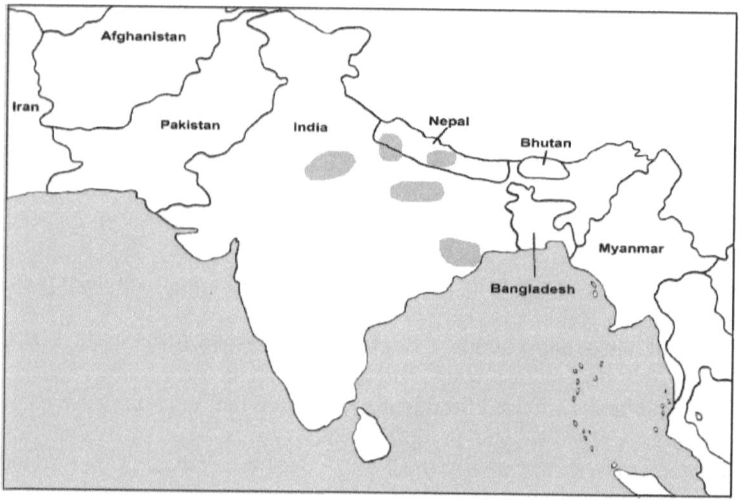

Source: Stevenson, C. and Whitaker, R. (2010)

This book reflects the overall the status of gharial conservation in Narayani River. It also deals with the information on various aspects such as the habitat requirements, population status, nesting ecology, threats and conservation and management aspects.

1.1 Morphological characteristics

The gharial is characterized by its extremely long, thin jaws, regarded as an adaptation to a predominantly fish diet.

The species is the only member of the Family Gavialidae, although recent molecular evidence suggests that *Tomistoma schlegelii* also belongs to this family (Densmore ,1983; Willis *et al.* 2007). The Gharial is the most thoroughly aquatic of the extant crocodilians, and adults apparently do not have the ability to walk in a semi-upright stance as other crocodilians do (Bustard & Singh 1978; Whitaker & Basu 1983).

It is dark or light olive above with dark cross-bands and speckling on the head, body, and tail. Dorsal surfaces become dark, almost gray-black, at about 20 years of age. The Ventrals are yellowish-white. The neck is elongated and thick. The dorsals are more or less restricted to the median regions of the back. The fingers are extremely short and thickly emarginated with a web. Males develop a hollow bulbous nasal protuberance at sexual maturity. The bulbous growth is used to modify and amplify "hisses" snorted through the underlying nostrils. The resultant sound can be heard for nearly a kilometer on a still day. Although the function of the bulbous narial excrescence is not well understood, it is apparently used as a visual sex indicator, as a sound resonator, or for bubbling or other associated sexual behaviors.

The average size of mature gharials is 3.5 to 4.5 m. The maximum recorded size is 6.25 m. The hatchlings have the size of approximately 37 cm. Young gharials can reach a length of 1 m eighteen months.

The average body weight ranges from 159 to 250 kg. Males commonly attain a total length of 3 to 5 m, while females are smaller and reach a body length of up to 2.7 to 3.75 m.

The elongated, narrow snout is lined by 110 sharp interdigitated teeth, and becomes proportionally shorter and thicker as animal ages. The well-developed laterally flattened tail and webbed rear feet provide tremendous maneuverability in deep water. On land, however, an adult gharial can only push itself forward and slide on its belly. The laterally compressed tail serves both to propel the animal and as a base from which to strike at prey. There are 27 to 29 upper and 25 or 26 lower teeth on each side. The length of the snout is 3.5 (in adults) to 5.5 times (in young) the breadth of the snout's base. Nuchal and dorsal scutes form a single continuous shield composed of 21 or 22 transverse series. Gharials have an outer row of soft, smooth, or feebly keeled scutes in addition to the bony dorsal scutes. They also have two small post-occipital scutes. The outer toes are two-thirds webbed, while the middle toe is only one-third webbed. They have a strong crest on the outer edge of the forearm, leg, and foot. Typically, adult gharials have a dark olive colour tone while young ones are pale olive, with dark brown spots or cross-bands.

The only breeding male in Narayani River of Chitwan National Park

Adult female gharial in the Gharial Breeding Center

1.2 Feeding habits

The elongated, narrow snout reduces resistance to water when snagging fish and the very sharp interdigitated teeth are well adapted for capturing fish. Young gharials eat <u>insects</u>, <u>tadpoles</u>, small fish and <u>frogs</u>. Adults feed on fish and small <u>crustaceans</u>. Their jaws are too thin and delicate to grab larger prey, especially a person. They catch fish by lying in wait for fish to swim by, and then catch the fish by quickly whipping their head sideways and grabbing it in their jaws. They herd fish with their body against the shore, and stun fish using their underwater jaw clap. They do not chew their prey but swallow it whole.

1.3 Nesting habits

The females reach sexual maturity at around 3 m total length. Nesting occurs during the annual dry season in hole nests excavated in river sand banks (Whitaker and Basu 1983). Clutch size is 30-50 eggs, and the eggs are the largest of any crocodilian (average 160 g).

In Chitwan National Park, the mating of gharials occurs in the months of December and January. After mating, the breeding females for nesting move in search of safe and suitable places. These are normally on the elevated sand banks without any human disturbances. In Narayani River, the nests of gharials are primarily found in Khoria Muhan and Velaunji.

A hatchling emerging out of the egg

1.4 Status of gharial distribution pattern

Before 1960, the gharials were abundant in the Narayani River and the adjoining Rapti River in the Chitwan National Park. The Chisapani Gorge, in the Bardia National Park also provided suitable habitat for the gharials. The construction of a bridge over the Karnali River in 1975 led to the degradation of gharial habitats. The gharial population in the Karnali, which numbered 15-20 adults in the mid-1970s, currently (1988) is estimated at 7 animals.

Hundreds of gharials were observed on the lower Narayani River prior to the construction of the dam on the river near the Indo-Nepalese border in 1964. In the early 1950s, about 235 gharials occurred along the river between Narayanghat and Tribeni. These gharials were concentrated mainly near sites, e.g., Litteguintha,

Kathauna, Velaunji, Narsahi, Khoria and Pitaungighat, that exhibited deep water and prevalent sand banks. Also in the early 1950s, there were 40-50 gharials near the border of Nepal and India: these gharials have been extirpated by Indian poachers by 1960. Approximately 75 wild gharials were extant in Nepal; in 1988. This population represents 16% of the world population. The largest single population, consisting of about 60 adults, was found on the Narayani River. Surveys between 1980 and 1987 revealed 51-60 wild gharials in the Kali and Narayani Rivers. Additionally, 307 captive reared gharials have been released in the Narayani, Kali, Koshi, and Rapti Rivers of Nepal since 1981.

Gharials use the Rapti and Gundrihi Rivers only during monsoon flood.

The gharial crocodile (*Gavialis gangeticus*) is a globally threatened and endangered reptilian species of Nepal. It is included as endangered in IUCN red list category and Appendix I by CITES. The natural population of gharial in the world is estimated at about 200 of which remnant population of about 80 was estimated recently by the government of Nepal. The gharials are found in the Narayani, Rapti, Babai and Karnali rivers of Nepal which are under tremendous threats from human disturbances such as overfishing, grazing, dam construction and over-exploitation of natural resources.

Nepal's rivers can be divided into three categories in accordance with their origins. The first category comprises the three main river systems of the country-the Koshi, Gandaki and Karnali river systems, all of them originating from glaciers and snow-fed lakes.

The Koshi river system consists of the Tamor, Arun, Dudhkoshi, Likhu, Tamakoshi, Sunkoshi and Indravati rivers. Of these, the Arun and Sunkoshi originate in Tibet. The confluence of these rivers is at Tribeni in Sagarmatha Zone. Flowing for almost 10kms through a narrow gorge before entering the plains, the "Sapta Kosi" or he "Koshi" swollen with the waters of the seven rivers finally merges into the Ganges.

The Gandaki river system in central Nepal consists of the Kaligandaki, Budhigandaki, Marsyanghi, Trishuli, Seti, Madi and Daraundi rivers. The Kaligandaki is the longest river and the Trishuli, the main tributary of this system.

The Kaligandaki originates in Mustang and converges with the Trishuli at Deoghat in Chitwan. The river is then called the Narayani and goes on to meet the Ganges. The Karnali river system in western Nepal consists of the Humla Karnali, Mugu Karnali, Seti and Bheri rivers and is the longest river system in the country. The Humla Karnali, which rises in Tibet, is the main tributary. After entering India, this river assumes the name Gogra.

The rivers like the Mechi, Mahakali, Bagmati, Kamala, Rapti, etc., most of which have their origin in the Mahabharat range, constitute the rivers of the second category.

The gharials in Nepal are distributed in Rapti, Narayani, Karnali and Babai river systems. Previously, they were found in the Koshi River, but have been extirpated due to the habitat loss and high human disturbances. The recent population monitoring census carried out by Chitwan national Park revealed 42 gharials in Narayani, 24in Rapti, 10 in Karnali and 9 in Babai rivers. These data suggests that largest population are found in Narayani and Rapti due to the regular monitoring and effective conservation measures given by park officials as compared to other river systems.

1.5 Gharial Conservation Breeding Center

In Nepal, the gharial conservation project was started in 1978 added by the Frankfurt Zoological Society. Gharial in Narayani and Kali Rivers have increased from 53 in 1981 to about 90 in 1987 by restocking captive reared gharials into the rivers. Since 1981 the gharial conservation project adopted several strategies which operated simultaneously in the Chitwan National Park. They included updating the status of wild gharials in Nepal, identifying suitable habitat for reintroduction and protection, collection of wild eggs from the Kali and Narayani Rivers for hatchery incubation and rearing at Kasara, reintroduction of captive reared stock, and long term monitoring of the effectiveness of reintroduction. The number of reared gharials that have been released till 2007 in the different rivers of Nepal is 691 individuals (Table 1).

The project also provided eggs and live animals to a variety of zoos and crocodile rearing centers in Japan and the United States for captive breeding programs.

The Nepal Gharial Conservation Project in the past has successfully produced over 1500 gharials and reintroduced 307 into the Narayani, Kali and Koshi Rivers.

Table 1: Number of Gharials Released after breeding in Captivity

S.N.	Release Year	Narayani	Rapti	Kali Gandaki	Sapta Koshi	Karnali	Babai	Total
1	1981	50	0	0	0	0	0	50
2	1982	50	0	0	0	0	0	50
3	1983	25	0	35	42	0	0	102
4	1984	15	0	0	0	0	0	15
5	1985	0	5	0	0	0	0	5
6	1986	0	0	0	43	0	0	43
7	1987	43	0	0	0	0	0	43
8	1988	0	0	0	0	0	0	0
9	1989	0	0	0	0	0	0	0
10	1990	25	0	0	0	0	30	55
11	1991	0	0	0	0	0	20	20
12	1992	38	0	0	0	20	0	58
13	1993	5	0	0	0	0	0	5
14	1994	0	0	0	0	0	0	0
15	1995	27	0	0	0	3	0	30
16	1996	19	0	0	0	0	0	19
17	1997	10	0	0	0	0	0	10
18	1998	15	5	0	0	0	0	20
19	1999	0	7	0	0	0	0	7
20	2000	7	0	0	0	0	0	7
21	2001	0	0	0	0	0	0	0
22	2002	10	0	0	0	0	0	10
23	2003	36	0	0	0	0	0	36
24	2004	0	20	0	0	0	0	20
25	2005	0	10	0	0	0	0	10
26	2006	0	20	0	0	0	0	20
27	2007	24	32	0	0	0	0	56
Total		399	99	35	85	23	50	691

Source: DNPWC 2007

2.0 Rationale of the study

The gharial is one of the key predators of the Narayani river system which helps in maintaining the health of the river and biotic integrity. The species is protected by Nepal Government under the National Parks and Wildlife Conservation Act 1973. The population of gharial crocodile in the Narayani is estimated at 44 (DNPWC, 2011). This remnant fragmented population is under continuous pressure from different human activities. On the other hand, the construction of Gandak barrage along the international border with India has further threatened the population through rapid loss of habitats. The government of Nepal has given priority towards the conservation of gharial. Realizing the declining trend of population and loss of habitat of gharial, the government has formulated the Gharial Conservation Action Plan (DNPWC, 2012). However, detailed research studies on habitat requirement, population, reproductive success and nesting and hatching behavior are lacking, which may be one of the constraint in gharial conservation.

This research has to some extent derived valuable information on habitat requirements and hatchling success which could contribute in designing the effective conservation strategies to maintain long term conservation of gharials in the river basins of Nepal.

3.0 Literature review

Andrews and Whitaker (2004) reports that *Gavialis gangeticus* females attain sexual maturity when they are about 3 m and males 3.5 m and above. The nesting season is during summer occurring between March and May and the maximum number of nests was laid between the 16th to 20th April. Total clutch weight highly correlated with clutch size than with mean egg weight. Significant variation in clutch size and percentage of viable eggs within a nest was evident between nesting seasons in the 13 years of breeding. Age and clutch size were found to be highly correlated and total clutch weight is strongly correlated with mean clutch size. *G. gangeticus* has temperature- dependent sex determination (TSD) and the TSD pattern is female-male-female. Fertile eggs incubated at set constant temperatures from 29-31.5°C produced only females and at 32°C 89% were males. At 33 and 33.5°C, 20% and 15% males were produced, respectively. Incubation period averaged 70 days for eggs incubated at 31°C,which was is 1.20 times higher than eggs from 29°C and 1.17 times longer than 33°C (Lang and Andrews 1994).

Khadka *et al* (2008) have conducted a census that revealed a minimum of 58 wild and released gharials are surviving in Chitwan population with 34 individuals in Narayani and 24 individuals in Rapti Rivers. TheKarnali and Babai rivers of Bardia National Park support a very small population of 6and 10 individuals respectively. The gharial is extinct from the Koshi River.The sex ratio of Chitwan population is only 1 Male: 6 Female and existence of only 27 percent of adult breeding population poses hindrance for its natural multiplication. The factors responsible for the decrease of gharial population are due to flood and dam construction, habitat destruction and decline in food. Overfishing, use of gill nets and river poisoning has also adversely affected the gharial population. The killing of gharial was only found to take place in case the animal got entrapped in the fishing net. The study also revealed high awareness among local community regarding gharial conservation as it has benefited local communities through the conservation programs associated with the species. The study concluded the declining population of gharial should be addressed through scientific study and preparing a Gharial Conservation Action Plan for regulating conservation activities to conserve gharial in Nepal.

Maskey *et al* (2006) reports that the captive breeding at the Gharial Conservation Project launched since 1981 is successful but survival of the released animals is very low. The recent observation of the gharial in the Narayani and Rapti rivers indicated that the population of the adult gharialis declining but it is compensated by the regular release of the captive reared animals though the survival is very low. As part of reinforcement program, activities such as construction of scientifically improved hatchling pools and regular monitoring of the released gharials in the Narayani River have been launched by the Department of National Parks and Wildlife Conservation.

Maskey *et al* (1995) have evaluated habitat of gharials in June, 1987 and January, 1988 between Sikrauli and Tribenighat, surveying channels on each side of large, wooded islands for a total distance of about 180 km.

Thus, use of all 1778 sightings would greatly inflate sample size; therefore, for all analyses we scaled our habitat-use observations to include only single sightings of the 50 wild, adult gharials known to live in the study area. Although gharials used all habitats, we did not observe them across the five habitat types in proportion to habitat-type abundance in either sea-son (monsoon: $X2 = 171.51$; $df = 4$; $P < 0.01$; non-monsoon $X2 = 85.89$; $df = 4$; $P < 0.01$) (Table 1). During monsoon gharials frequented sand banks ($X2 = 163.86$; $P < 0.01$; df throughout this paragraph = 1) and avoided grass banks ($X2 = 44.18$; $P < 0.01$); during non-monsoon season animals selected sand banks again ($X2 = 53.50$; $P < 0.01$) and avoided rock banks ($X2 = 54.63$; $P < 0.01$). We could not definitively demonstrate other habitat preferences or avoidances. Even though habitat-type availability changed substantially between summer and winter, proportional habitat-type use remained rather constant. This trans-seasonal consistency in observed habitat use, despite changing proportions of habitat availability, suggested that general site fidelity might be important in defining habitat use. However, long-term site fidelity cannot be the sole factor at work. Observations (N

= 1474) on 43 newly-released, captive-reared gharials (1-2 m total length) showed the same pattern of habitat-type preference and habitat-type avoidance; furthermore, these habitat-use patterns persisted even when the animals moved long distances from original release sites (Maskey, 1989). Habitat selection may be mediated (1) by tactile qualities of substrate, (2) thermoregulatory considerations, and/or (3) by prey availability. Data do not presently exist to test any such hypotheses.

Maskey (1999) conducted a field study of gharial was conducted in the Chitwan National Park and Bardia National Park during 1997 to determine the status of gharial in the Kali, Narayani, Karnali and Babai river systems of Nepal. Systematic survey conducted in December revealed a minimum of 55 wild gharials and 50 released gharials survived in the Narayani, Kali, Karnali and Babai rivers. The sex ratio of wild gharial 1 male to 10 females was at critical stage. The low number of males was attributed to the heavy poaching of males in the past. The population may be sustained by releasing captive-released gharials.

Maskey and Percival (1994) conducted a study in the Royal Chitwan National Park and Royal Bardia National Park during 1993 to determine the status of gharial in the Narayani, Kali, Karnali, and Babai river systems of Nepal. Systematic survey conducted in December and May revealed that a minimum of 58 wild gharials and about 70 released gharials survived in the Narayani, Kali, Babai, and Karnali rivers. The sex ratio of wild gharials was 1 male to 10 females. The low number of males was attributed to the heavy poaching of males in the past. The population may be sustained by releasing captive-reared gharials.

Nair (2010) studied the gharial in a 75 km stretch of Chambal River and concluded that one-fifth of the study area as preferred gharial habitat. The availability of undisturbed basking sites in conjunction with deep water segments emerged as the main variable explaining gharial occurrence. The human activities appeared to negatively influence the use of areas by gharials. The mining of sand and cultivation around the banks negatively impacted the use of such sites for basking. The gharials were seen less often and in fewer numbers in areas where fishing was high. Similar results were seen with movement of people and livestock along the river stretch. This study indicates the importance of inviolate areas that satisfy the bio-physical requirements of the gharial.

Nair *et al* (2012) surveyed 75 km section of the river Chambal and photographed individual gharials for capture–recapture analysis. The total sampling effort yielded 400 captures. Used within the framework of capture–recapture analysis, photo identification provided a reliable and noninvasive method of estimating population size and structure in crocodilians. They also opined that without determining the current status of gharials, highly intensive strategies, such as the egg-collection and rear-and-release programs that are being implemented currently, initiated on the basis of underestimates of population sizes, are unwarranted and divert valuable conservation resources away from field-based protection measures, which are essential in the face of threats like hydrologic diversions, sand mining, fishing and bank side cultivation.

Hussain (1999) studied the reproductive success and hatchling survival of gharial populations in Chambal Sanctuary by monitoring 124 nests between 1987 and 1989. The population increase in the Chambal River of the Sanctuary was also determined between 1988 and 1992. The study showed that between egg laying and hatchling, a large number of eggs were lost due to eggs being damaged during nest searches, predation, desiccation and other reasons although the overall fertility was 91.8%. The study indicated that 92.3% hatchling mortality occurred within the first year. The density of gharial increased during 1992 and the total number of nests also increased in 1989.

A population survey by Thapaliya *et al* (2009) carried out in the river systems of Nepal indicated the presence of 70 gharials. The study showed decreasing trend of gharial population due to high human disturbances. The study identified a need to carry out scientific study based conservation measures such as restocking and habitat improvement.

Hussain (2009) studied the basking site and water depth selection by gharials in National Chambal Sanctuary between 1992 and 2007 showed the greater habitat preferences to sandy parts of the river banks and sand bars for basking. The juvenile gharials preferred the water depth from 1-4 m while the adult and sub-adult prefer the deeper water greater than 4 m. Sand excavation and water abstraction are the major threats to gharials in Chambal Sanctuary.

Khadka & Thapaliya (2010) surveyed by the use of radio-telemetry in Narayani and Rapti rivers to assess the habitat preference on the basis of degree of human disturbance showed that the gharials preferred less disturbed area The study also estimated the population of 56 wild and released gharials in Chitwan Natioanl Park which includes 36 individuals in Narayani and 20 individuals in Rapti Rivers

Maskey (1989) conducted a field survey of gharial in Chitwan National Park from 1986-1987 to determine its status and ecology in Narayani River. The survey during the months of December and May indicated a minimum of 103 (0.93 gharials/km) wild and released gharials, the sex ratio being 1 male:9 females. The habitat surveys of June 1987 and January 1988 showed that rocky banks were the most available habitat in the Narayani River, followed by sand banks, grass banks, sand grass banks and river channel. The larger gharials used sand banks more than on rocky banks. On the other hand, the small gharials used rocky banks more than the larger animals. A mean clutch size of 35.2 ± 1.1 SD resulted from 73 nests collected in the years 1977 and 1987. The mean incubation period and the hatching rates of eggs from the Kali and Narayani rivers were 78 ± 9 SD days and 60.9 %, and 81 ± 11.5 SD days and 67.7 % respectively. 25% to 80% of the hatchlings died within the first year of their life. The prime criteria for release of captive gharial are deep, fast-flowing clear water, high banks, deep pools and undisturbed sand banks at the river edge. The success of gharial reintroduction program in Nepal depends upon three factors: the release of gharials>1.2 m in length; the selection of release sites that provide primary habitats within the dispersion range of the released gharials, and release of gharials in late winter to facilitate the establishment of site-fidelity bonds by individuals prior to the monsoon season.

4.0 Policies and Legistations

Nepal is a signatory of over 25 major international convention and treaties, including United Nations Convention on Biological Diversity (CBD) of 1992, Convention on Wetlands of International Importance especially as Waterfowl Habitat called Ramsar Convention of 1971, World Heritage Convention of 1972, Convention on International Trade in Endangered Species of Wild Flora and Fauna (CITES) 1975, UN Framework Convention on Climate Change of 1997 and UN Convention to Combat Desertification of 1994.

4.1 Ramsar Convention

The Convention on Wetlands of International Importance especially as Waterfowl Habitat, known as the Ramsar Convention, was signed in 1971 and came into force in 1975. It is an independent international convention designed to protect the wetland ecosystems from further destruction. It calls on all signatories to conserve wetlands, promote their sustainable utilization, and set aside special areas as wildlife reserve. Every country is required to designate at least one wetland for inclusion on the list of wetlands.

The Government of Nepal ratified the Ramsar Convention in 1987, and designated Koshi Tappu Wildlife Reserve (KTWLR) for inclusion in the Ramsar site. Though Koshi Tappu Wildlife Reserve in the first Ramsar site of Nepal and three other wetland sites—Ghodaghodi Lake area, Beeshazar Lake, and Jagdishpur Reservoir have been recently nominated. These recent nominations all fall outside protected areas, and the legal basis for their conservation and management arrangements have not been clarified. Although the DNPWC is the focal point for Ramsar and is responsible for all Ramsar Sites, the Department of Irrigation and the Department of Forests respectively are responsible.

4.2 The Nepal Biodiversity Strategy, 2002

The Nepal Biodiversity Strategy (NBS) lays down Nepal's strategy for biodiversity conservation and has clearly identified the need for conservation and sustainable-use of wetlands. The NBS recognizes the need for a comprehensive approach aimed at conserving forests, soil, water, and biological diversity while at the same time meeting the basic needs of people who are dependent on these resources for their livelihoods through consolidation and continuation of past successful efforts. Cross-sectoral strategies of the NBS include landscape planning, integrating local participation, institutional strengthening, in-situ conservation, strengthening the National Biodiversity Unit etc. The Nepal Government has prepared the NBS Implementation Plan (NBSIP) to translate NBS visions into actions by addressing, among others, the issues related to management planning resources allocation and capacity development.

4.3 National Wetland Policy (2003)

In order to meet government's obligations under Article 3 of the Ramsar Convention to develop a national wetland policy, and under Recommendation 6.3 of the Conference of the Parties 1996, to mange wetlands in participation with local people and communities, the National Wetland Policy (2003) has been formulated. The primary goal of the National Wetland Policy is to conserve and manage wetland resources wisely and in a sustainable way with local peoples' participation. The policy also aims to put the conservation and management aspects of wetlands conservation within the framework of broader environmental management

The major objective of the policy is to involve local people in the management of Nepal's Wetlands and conserve wetlands biodiversity with wise use of wetland resources. The policy addresses the need for a coordinated approach to wetland management and includes the following objectives:

i. Identify Nepal's wetlands and prepare detailed management plans for each of them to prevent degradation and disappearances of wetlands.

ii. Identify local peoples' knowledge, skill and practice regarding wetlands and promote their innovations and traditional research for the sustainable use of wetland resources.

iii. Conserve and manage wetlands according to the needs and on the basis of scientific knowledge and technology.

iv. Promote Women's participation for the conservation, management and wise use of wetlands.

v. Gradually implement international treaties for wetland conservation.

vi. Disseminate information to raise public awareness about wetlands.

To fulfill these objectives following eight major policies on wetland has been worked out:

i. Wetland Management Policy based on Local Participation.

ii. Classification of wetland and Management Policy

iii. Policy regarding the wise use of wetlands

iv. Policy regarding the Promotion of Awareness.

v. Prevention, Control and Management of Invasive Species.

vi. Institutional Policy regarding Wetland Management

vii. Policy regarding Prohibition of works with adverse impact on Wetlands.

viii. Policy regarding disappeared and disappearing Wetlands.

The National Wetland Policy (2003) proposes Immediate and future works in line with these policies. Preparation of wetland inventories; formulation of related acts, regulation, guidelines; identification of wetlands eligible to be listed for Ramsar Site; financial arrangements, legal initiatives for intellectual property are the immediate works proposed by the policy. Similarly, activities such as international publicity, prioritization of wetlands, development of model wetland sites, minimization of the degradation of water sources, regulation of underground water and water pollution, organization development, management of trans-boarder wetland etc fall under the future works of wetland conservation and management policy.

4.4 Aquatic Animals Protection Act 1961 (AAPA, 1961)

The AAPA promulgated for protecting aquatic animals in natural water bodies like rivers, reservoirs and lakes has remained virtually defunct due to the lack of related bylaws/regulations. After 38 years of its promulgation, parliament revised it in 1998 to activate it. Section 5a included in this amendment permits only the use of safe pesticides in case any poisonous material is to be used for catching aquatic life. Sections (4a), (4b) and (5) empower the government to prohibit catching, killing and harming certain kinds of aquatic animals in different scenarios. Due to insufficient inter-ministry and inter ministry coordination and pending adoption of required regulation its enforcement is still pending. The District Development Committee (DDC), exercising the licensing of fishing in natural water bodies has not been given responsibility of aquatic life protection. Similarly, the Ministry of Water Resources (MoWR), the authority of the natural water bodies is not empowered legally for the conservation and protection of aquatic life.

4.5 The National Parks and Wildlife Conservation Act (NPWCA, 1973)

The *National Parks and Wildlife Conservation Act* (1973) laid the outline for the conservation of wildlife in the country and lists protected species. This list has been amended four times since 1973 (1974, 1982, 1989 and 1991), with inadequate inclusion of globally threatened species. According to this Act number of area of the country are declared as Protected Areas (PAs), National Parks, Wildlife Reserves, Conservation Area, Hunting Area, Buffer Zones etc. The DNPWC presently works with a network of eight National Parks, four Wildlife Reserves, three Conservation Areas, one Hunting Reserve which, including five buffer zones (including Bishazari Lake area) around national parks, cover a total of 26, 696km^2 or 18.14 percent of the country's total land area and 67.89 percent of ecosystems of the country. The Ministry of Forests and Soil Conservation (MFSC) approved the Buffer Zone Management Guidelines in 1999. The Guidelines have been implemented in the five Buffer Zones already declared by HMG/Nepal. Detailed guidelines have been provided for the implementation of provisions related to the

Buffer Zones of the National Parks and Wildlife Conservation Act as well as the Buffer Zone Management Regulation at the field level. Moreover, it facilitates the work of government staff and User Committees in Buffer Zone Programmes. In addition, regulation for government-managed conservation areas has been passed to enhance community participation in conservation and local development.

5.0 Objectives

The overall objectives of the study is to strengthen the gharial conservation in Narayani River, Chitwan National Park. The specific objectives are: 1) assess the population status of *Gavialis gangeticus* in Narayani River; 2) assess the habitat utilization by gharials; 3) prepare GIS maps of distribution and habitat; 4) raise the level of awareness on gharial conservation among the local communities.

6.0 Study Area

The study was carried out in Narayani river of Chitwan National park (27^0 34' to 27^0 68' N and 83^0 87' to 84^0 74' E) including the buffer zones from northern boundary of park (Sikrauli) to Tribeni barrage at international border with India. Chitwan National Park is renowned for the conservation of some of the world's most endangered species, including rhinoceros, tiger, gaur, gangetic dolphin, otter and gharial in their natural habitat (CNP Management Plan 1975- 1979). The Park occupies an area of 932 km² in the Rapti Valley of the Siwalik physiographic region, while the buffer zone (27° 28' N to 27° 70' N and 83° 83' E to 84° 77' E) extends 750 km² area (Fig. 2).

Figure 2: Narayani River

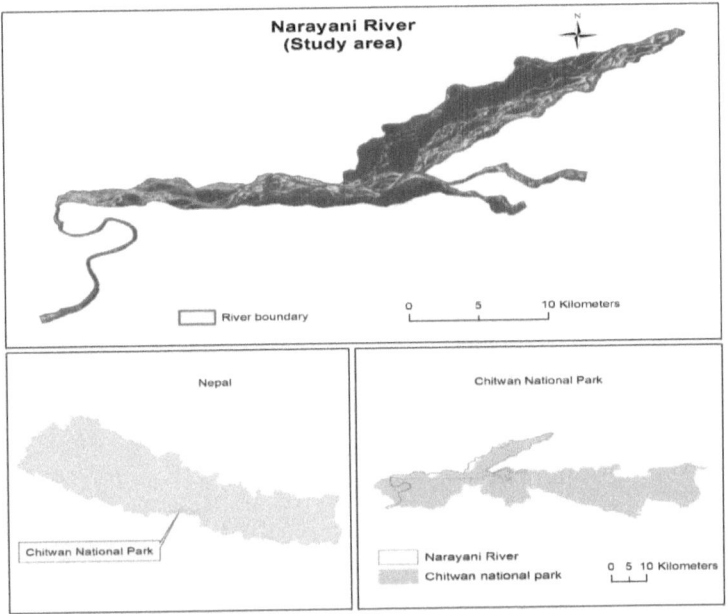

The Narayani River ((also called the Gandaki} is a snow fed river, formed by the confluence of Kaligandaki and Trisuli rivers. The total length of this river is about 338 km, and the average flow ranges between 1000 to 1700 cum/s but maximum flow ranges from 10 to 700,

000 cum/s during the monsoon season from June to September (Panday, 1987; Maskey, 1989).

The Narayani river flows southwest for 30 kilometers from a gorge in the Mahabharat Range to the Rapti confluence and then flows westward for a further 25 kilometers along the base of the Someswar Hills before turning south through a very narrow gorge in the Siwaliks between the Dauney and Someswar Hills until it reaches the Nepal-India border (Laurie, 1978). The bed of the Narayani River is very broad consisting of a large number of channels and islands with a width of up to 4 kilometers. The floodplain varies with the altitude, ranging from 250 meters to 150 meters.

The climate of Chitwan is subtropical with a summer monsoon from mid-June to late September and a relatively dry winter. The average annual rainfall is about 250 cm, with the most occurring between June and September. The post-monsoon season between November and January is cool with the daily average temperature reaching 24 ^{0}C during the day and droppings to about 7 ^{0}C at night.

7.0 Materials and methods

Research design

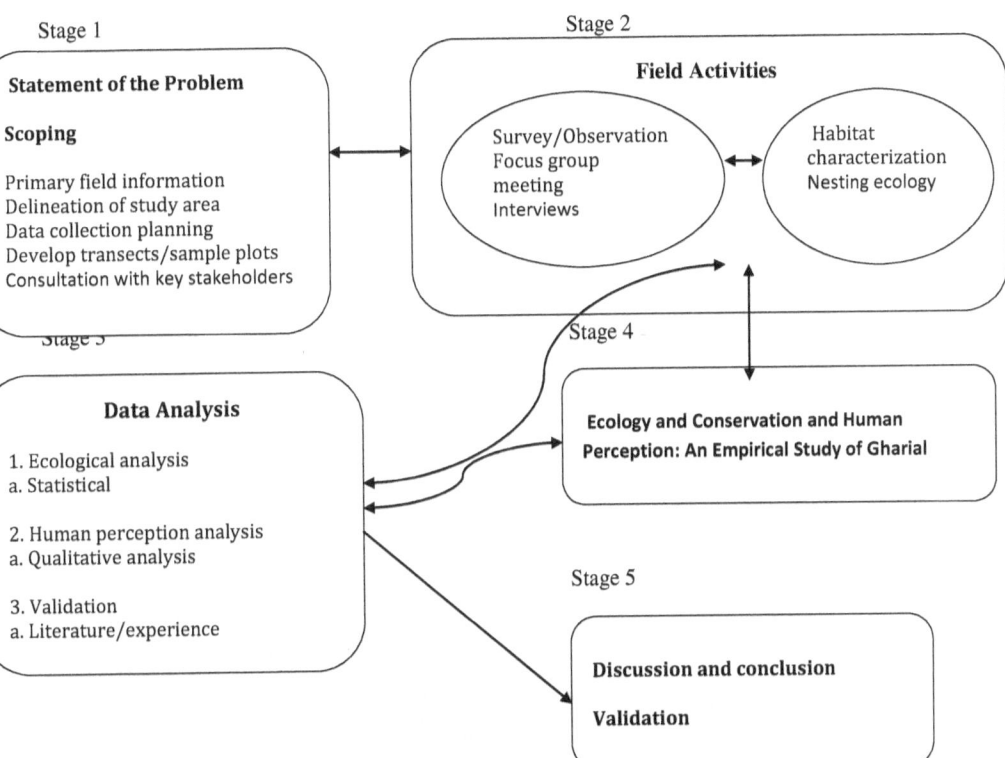

7.1 Scoping

The scoping was conducted prior to embarking a detailed field study to collect the primary field information.

7.2 Field activities

7.2.1 Survey

The survey of gharial was conducted in January 2013 from Rapti-Narayani confluence area to Tribeni barrage along Nepal-India border.

The survey was conducted in two dugout canoes with four fishermen. The details of the gharial sighted with the help of Olympus binoculars were recorded with their habitat features and GPS location (Annex 1).

Figure 3: Gharial survey locations - GPS points

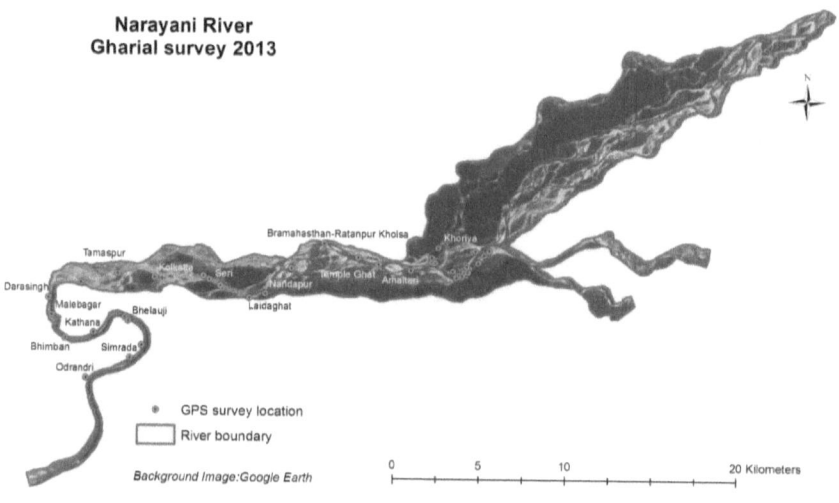

The gharials were classified into different size classes in 30 cm increments. The individuals < 120 cm long were considered as hatchlings, > 120-180 cm as juveniles, > 180-270 cm as sub-adults and those > 270 cm as adults (Hussain, 2009).

The sex of the sighted crocodile was also determined by presence or absence of nasal orifice.

7.2.2 Distribution and Abundance

The census of gharial was conducted to evaluate the population abundance in the different blocks of the study area. The details of the sighted gharials such as substrate type, river width; mid-river depth and degree of human disturbances were estimated.

Figure 4: Sightings of gharials

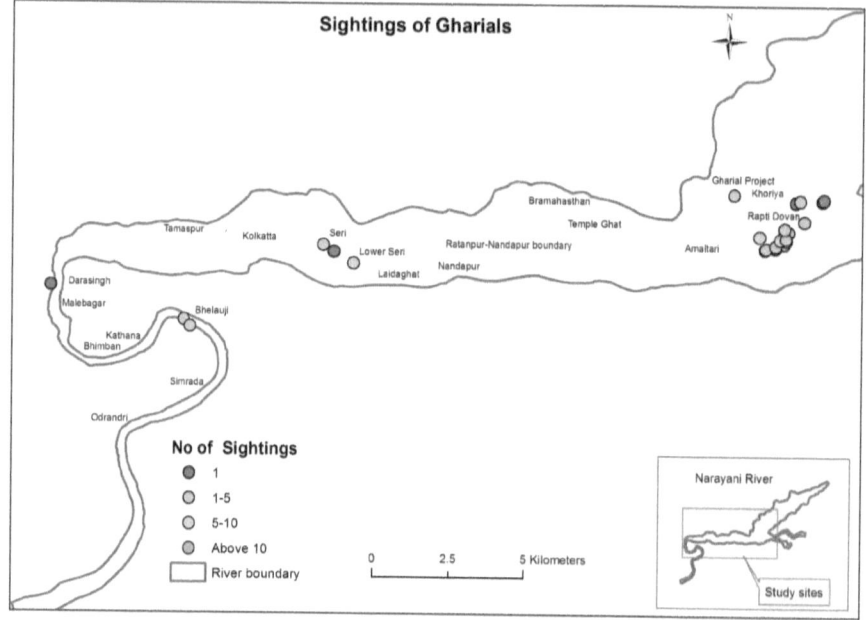

7.2.3 Basking

The study of basking activity was conducted in the two areas, namely Khoriya Muhan and Velaunji for three and two days respectively. The details of the sighted gharials on both banks of the river as well as the mid-sandbars and islands were noted. These include approximate size, basking site topography, substrate characteristics and the mid-river water depth and width.

7.2.4 Hatching success

Hatching success of the artificially incubated eggs in Gharial Project were studied. The details of the eggs such as the number of eggs hatched, eggs damaged, and hatchling mortality in each clutch collected from different nests were recorded.

7.2.5 GIS study

The field data such as GPS locations of sign survey and riverine habitat features have been visually interpreted from the Google Image 2010 after ground truthing. The data were generated by the screen digitization in Arc GIS 10.0 version.

7.2.6 Awareness Raising

The awareness materials in the form of brochure was prepared in the native language and distributed widely among the school students adjoining the park area, buffer zone communities and information centers of the national park. This will promote awareness and develop stewardship towards conservation of aquatic fauna, particularly gharials among the different target groups.

The focal group/stakeholder meetings in collaboration with the park were conducted in Khoriya Muhan, Gharial Project Camp, Amaltari and Gharial Breeding Center, Kasara. The participants included the fishermen, nature guide, park staff and buffer zone communities representing different places adjoining the park area. These meetings were fruitful in developing strong linkage between stakeholders regarding the conservation of gharial. The meetings will also help in understanding people's perception and identifying conservation problems as well as suggestions for means of gharial conservation.

8.0 Result

8.1 Population status

A survey was carried out in the Narayani River from Narayani – Rapti confluence to Tribeni at the international border with India. The survey recorded a total number of 38 gharials. Among the observed gharials, 3 were hatchlings, 2 juveniles, 12 sub-adults and 15 adults. Only one breeding male was recorded. The abundance of the gharial population was mostly confined largely in three areas, namely, Khoriya, Seri and Bhelauji (Table 2 & 3). Among the gharials observed in these areas, Khoriya had the largest congregation and Seri had only few numbers of gharials.

Table 2: Gharial survey in the Narayani River, January 2013

River	Hatchlings	Juveniles	Sub-adults	Adults	Totals	Remarks
Narayani River	3	8	12	15	38	One breeding male
Total	3	8	12	15	38	

Table 3: Size classes of gharial seen in the Narayani River, CNP during 2013.

Size class of gharial	2013
<120 cm	3
>120 – 180 cm	8
>180 – 270 cm	12
>270	115
Total	38

Recent monitoring survey in Narayani and Rapti rivers carried out by Chitwan National Park in November 2012 recorded altogether 87 gharials, of which 52 were recorded in the 100 km stretch of Narayani River. The census of gharial according to the age classes had maximum of sub-adults representation with a total number of 28 individuals and hatchling represents a minimum number with only 4 individuals (Tables 4 & 5).

Table 4: Results of Gharial counts in the Rapti and Narayani Rivers, November 2012

River/Location of sightings	km	Hatchlings	Juveniles	Sub-adults	Adults (M.F)	Totals	Remarks
Rapti River	50.0	0	9	22	4 f	35	No male
Narayani River	100	4	6	28	14	52	One male
Grand Total	150	4	15	50	18	87	

Source: GBC, CNP, 2012

Table 5: Size classes of gharial sen in the Narayani River, CNP during 2012.

Size class of gharial	2012
<120 cm	4
>120 – 180 cm	6
>180 – 270 cm	28
>270	14
Total	52

Source: GBC, CNP, 2012

8.2 Habitat utilization (Basking sites)

In the basking areas, the water channels were from 1.2 m – 4.5 m deep. In Khoriya population, the gharials used the sand bars and rocks for basking with the river channels having 1.2 m – 2.7 m deep whereas in Bhelauji, the basking sites were only on sand with maximum water depth of 3.6 m (Figs. 5 & 6).

The field observation showed that more than 80 percent of gharials select sandy areas for basking activities followed by rocks with about 15 percent. No gharials were observed in clay (Table 6).

Table 6: Preference of basking types by gharials in Narayani River, CNP during January 2013

Basking site type	Number of gharial observed	Percentage of habitat use
Sandy	32	84.21
Rocky	6	15.78
Clay	0	0
Total	38	100

Figure 5: Habitat utilization by gharials

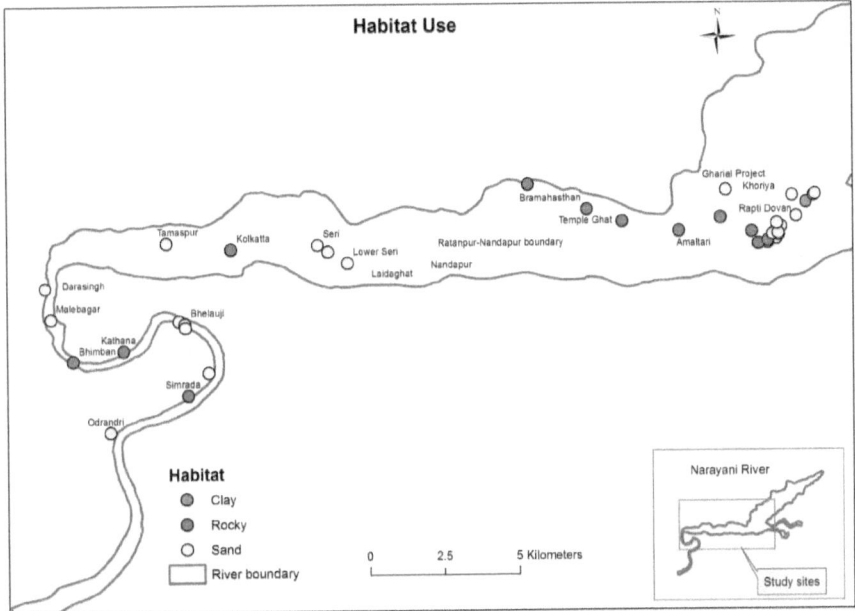

Figure 6: Water depth selection by gharials

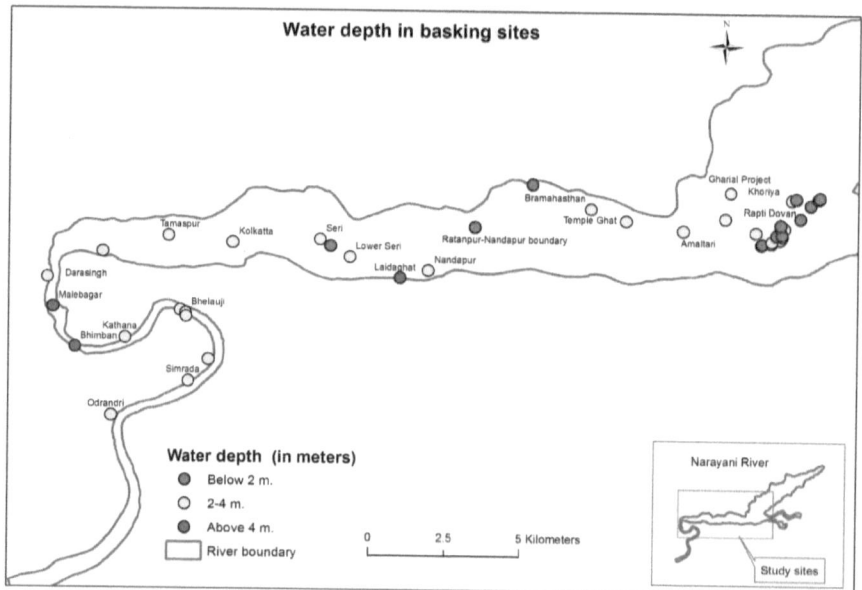

8.3 Hatching success

Table 7: Reproductive effort of gharial in Narayani River, CNP in 2012

S. No.	Egg laying date	Location	Clutch size	Hatched date	Incubation period	Hatchling no.	% of eggs hatched	Death during hatching	Post-hatching mortality	Infertile eggs
1	5 April	Khoriya Muhan	18	14 June	71	9	50.0	0	0	9
2	31 March	Hattisar, Khoriya	36	16 June	78	8	22.2	0	7	21
3	31 March	Bhelauji	41	16 June	NA	0	0.0	0	40	1
4	4 April	Bhelauji	33	15 June	71	12	36.3	0	19	2
5	5 April	Bhelauji	30	15 June	70	2	6.6	0	10	18
	Sub-Total		158			31	19.6	0	76	51

48

The average clutch size of the gharials that were collected during 2012 is 31.6 with a maximum clutch size 41 and a minimum of 18. The average percentage of eggs hatched is 19.6 with a maximum 50 percentage and a minimum 0 (Table 7). Almost 50 percent of the eggs that were hatched were lost due to death by various reasons.

The review of hatching success of gharial in the river system of CNP from the year 1977 to 2011 is presented in Annex 2.

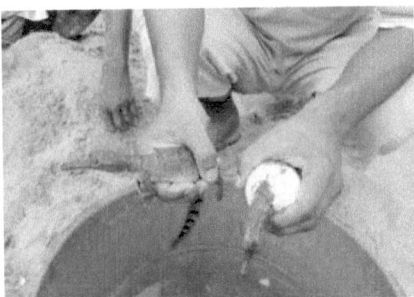
Gharial hatchlings in Gharial Project, Amaltari Island

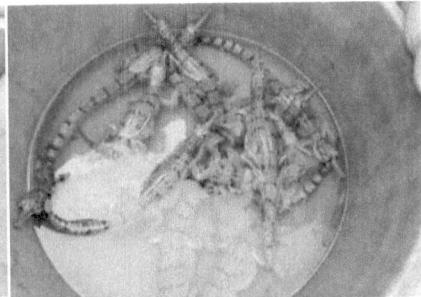
Gharial hatchlings from Gharial Project

Gharial eggs being collected from the natural nest

8.4 Threat assessment

Despite of all the conservation actions, gharial population in Nepal is staying at a critically low level. The major threats to the gharials are identified as:

Poaching: it has been clearly identified (Maskey, 1998) and the last observations show that poaching still exist in Nepal. It was mainly targeted on male gharials before.

Industrial pollution: The most widespread form of pollution is organic waste from domestic and industrial sources. There is a little doubt that pollution could be the cause of gharial population decline. Increasing industrialization is leading to increase in pollution loads from factories. Discharges from the Gorkha Brewery and Bhrikuti Paper and Pulp factory and Pharmaceutical and Gill Mary were the major sources of pollution in Narayani River affecting the gharial crocodiles.

Gandak barrage: The large dam in Tribeni has been built between Nepal and India that allow gharials to go in India following the stream, but once in the Indian side, they can't go back in Nepal. During the monsoon season, the huge stream bring a large number of gharial to India. Thus, it's impossible for them to return to their original habitat, decreasing the Nepalese population.

Overfishing: The Chitwan National Park have provided the fishing license to the traditional fishermen to support their livelihood. Besides, this wetland dependent communities, others are also intensively fishing in the river on both banks resulting to scarcity of fish prey base, disturbances to the gharials and loss of habitat. The fisher men use large fishing net (gill net) which largely threatens the gharial population due to risk of being entrapped. Small sized mesh nets are often used which removes both adult breeding stock and fingerlings from the populations reducing the possibilities of future breeding and recruitment from the areas.

Figure 7: Human disturbances in Narayani River

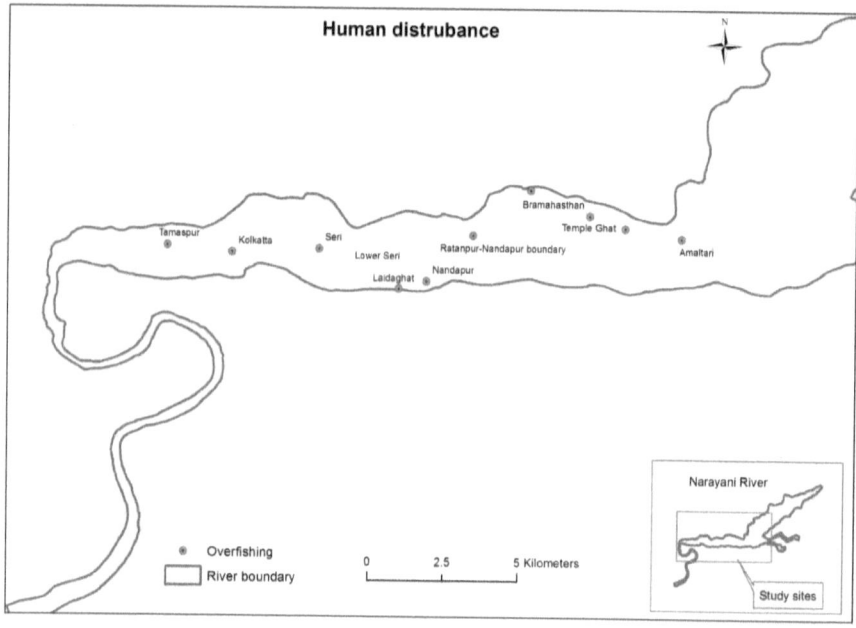

9.0 Discussion

The population census survey carried out in Narayani river showed the congregations of gharials, particularly in two sections of the river, Rapti confluence-Khoria Muhan and Velaungi in the downstream section. This study showed that less human disturbances because of restriction on fishing activities have resulted the presence of gharials in large numbers. In other sections of the river, the January 2012 survey did not sighted the gharials. This is due to the presence of high human disturbances such as fishing, grazing, extraction of sand and boulder and construction of Tribeni dam (Acharya & Rajbhandari, 2009; Acharya & Rajbhandari, 2012). Other studies also have indicated the threats to the gharial population by human disturbances and dam construction (Maskey, 1987; Collares *et al*, 2000). The gharials from different areas tend to migrate to the habitats with sand bars and sandy islands with deep water courses for safety and availability of its prey.

The presence of main channel, braided channels with sandy bars and patches of sandy islands within the main channels reflects the availability of sites for basking and food resources. These habitat features in Narayani and Rapti river confluences where the riverine habitat is characterized by main channel with braided channels, large areas of sand bars in between these channels, moderate to deep water and regular monitoring of habitats have tended the gharials ro select these areas during the winter season.

In the downstream areas such as Velaungi, the riverine habitat is characterized by the stagnant conditions of thenriver due to Tribeni dam. The river here flows in between the foothills of Mahabharat range and churia range. The presence of single deep water channels and occurrence of elevated sandy areas on the eastern banks with low human disturbances have caused gharials to prefer these sites. Additionally, these sites are important sites for breeding of gharials where eggs are collected annually for artificial hatching. This year, 3 nests of gharial eggs were collected from the Velaungi area (personal communication, Juthe Ram and Sante Bote, 2013).

Altogether 53 nests were collected from the Narayani River during the 6 nesting seasons of 1977-81 and 1987 (Maskey, 1987). During the study, only 5 nests with a total of 158 eggs were collected from from Velaungi, Khoria Muhan and Hattisar Khoria. The significant reduction in the number of nesting sites over years is due to high fishing pressures. The fishermen use large gill nets (Tiyari) for fishing which collect many small to large fish causing reduction in the prey base. Occasionally, the gharials get entangled in these nets.

10.0 Conclusion and recommendations

The gharials in Narayani are under immense pressure from anthropogenic disturbances and barrage construction. On the other hand, one of the major obstacles in the conservation of gharials in the river basins of Nepal is the wide gap in their researches. Such gaps have hindered the conservation due to the lack of basic information. The park should adopt appropriate conservation measures such as ban of fishing, promotion of prioritized researches and strengthen the policies to address the conservation of gharials so that their long term existence is ensured.

This study has suggested the following conservation measures to the improvement of gharial habitats and maintaining the viable population of gharial in Chitwan National Park:

- Conduct regular monitoring and patrolling of gharial habitats and population and ban the fishing activities in these areas.
- Conduct scientific researches on the gharials in Narayani on habitat requirements, home range and reproductive activities.
- The prime concern is the protection of a single breeding male in Narayani River. Use of radio-telemetry to monitor its movement and behavioral activity is necessary to ensure the maintenance of gene pool.

11.0 References

Acharya P.M. & Rajbhandari S.L.2012. Habitatn Ecology of *Lutrogale perspicillata* in Narayani River, Chitwan National park. In proceeding of Nepalese conference for Rufford grantees, Kathmandu, Nepal. 133-145 pp.

Acharya P.M. & Rajbhandari S.L. 2012. Investigation of population status and habitats of *Lutrogale perspicillata* in Narayani River, Chitwan National park. A second phase research report submitted to Rufford Foundation, U.K. 50 pp.

Andrews, H.V. & N. Whitaker. (2004). Captive Breeding and Reproductive Biology of the Indian Gharial Gavialis gangeticus (Gmelin).

Bustard, H.R. and Singh, L.A.K. (1978). Studies on the Indian gharial *Gavialis gangeticus* (Gmelin) (Reptilia, Crocodilia). Change in terrestrial locomotory pattern with age. J. Bombay Nat. Hist. Soc. 74: 534-536.

Collares-Pereira M.J., Cowx I., Ribeiro F., Rodrigues J. and Rogado L. 2000. Threats imposed by water resource development schemes on the conservation of endangered fish species in the Guadiana River basin in Portugal. *Fisheries Manage. Ecol.*, **7**: 167-178.

Densmore III, L.D. (1983). Biochemical and immunological systematics of the Order Crocodilia. Evolutionary Biology 16: 397-465.

DNPWC, 2007. Annual Report (Sharwan 2063 – Asadh 2064).

GCA (Gharial Conservation Alliance) (2008). International Gharial Recovery Action Plan. http://www.gharialconservation.org/PDF/GRAP.pdf.

Hussain, S. A. (1999). Reproductive success, hatchling survival and rate of increase of gharial (*Gavialis gangeticus*) in National Chambal Sanctuary, India. *Biological conservation*, 87(2), 261-268.

Hussain, S. A. 2009. Basking site and water depth selection by gharial Gavialis gangeticus Gmelin 1789 (Crocodylia, Reptilia) in National Chambal Sanctuary, India and its implication for river conservation. Aquatic Conservation-Marine and Freshwater Ecosystems 19:127-133.

Khadka, M., Kafley, H., & Thapaliya, B. P. (2008). Population Status and Distribution of Gharial (Gavialis gangeticus) in Nepal.

Khadka, M. & Thapaliya, B. P. (2010). Gharial (*Gavialis gangeticus*) Conservation Program.

Laurie, A. 1978. The ecology and behavior of the greater onehorned rhinoceros. Ph. D. Dissertation, Univ. Cambridge. 450 pp.

Martin, B.G.H. and Bellairs, A.D' A. (1977). The narial excresence and pterygoid bulla of the gharial, *Gavialis gangeticus* (Crocodilia). J. Zool. Lond. 182: 541-558.

Maskey, T.M. 1989, Movement and survival of captive –reared gharial, *Gavialis gangeticus* in the Narayani River, Nepal, A dissertation presented to the graduate school of the University of Florida in partial fulfillment of the Degree of Doctor of Philosophy.

Maskey, T. M., Percival, H. F. (1994). Status and conservation of gharial in Nepal. Submitted to 12th Working Meeting Crocodile Specialist Group, Pattaya, Thailand, May 1994.

Maskey, T. M., Percival, H. F., & Abercombie, C. L. (1995). Gharial habitat use in Nepal. *Journal of herpetology*, 29(3), 463-464.

Maskey TM. 1999. Status and conservation of gharial in Nepal. In ENVIS Bulletin on Wildlife and Protected Areas. Wildlife Institute of India: Dehra Dun, Vol. 2; 95–99.

Maskey, T., Cadi, A., Ballouard, J.M., & Fougeirol L. (2006). Gharial conservation in Nepal: Results of a population reinforcement program.

Nair, T. (2010). *Ecological and anthropogenic covariates influencing gharial Gavialis gangeticus distribution and habitat use in Chambal River, India* (Doctoral dissertation, Thesis submitted in partial fulfillment of Master of Science in Wildlife Biology and Conservation. Tata Institute of Fundamental Research, National Centre for Biological Sciences, India).

Nair, T., Thorbjarnarson, J. B., Aust, P., & Krishnaswamy, J. (2012). Rigorous gharial population estimation in the Chambal: implications for conservation and management of a globally threatened crocodilian. *Journal of Applied Ecology*.

Panday, R. K. 1987. Effect of Altitude on the Ge-ography of Nepal. Cent. Altitude Geography. Kathmandu.

Smith, M.A. (1931). Loricata, Testudines. In The Fauna of British India including Ceylon and Burma. Reptilia and Amphibia. Vol. I. Taylor and Francis: London.

Stevenson, C. and Whitaker, R. (2010). Indian Gharial *Gavialis gangeticus*. Pp. 139-143 *in* Crocodiles. Status Survey and Conservation Action Plan. Third Edition, ed. by S.C. Manolis and C. Stevenson. Crocodile Specialist Group: Darwin.

Thapaliya, B. P., Khadka, M., & Kafley, H. (2009). Population Status and Distribution of Gharial (Gavialis gangeticus) in Nepal. *The Initiation*, *3*, 1-11.

Willis, R.E., McAliley, L.R., Neeley, E.D. and Densmore III, L.D. (2007). Evidence for placing the False Gharial (*Tomistoma schlegelii*) into the family Gavialidae: Inferences from nuclear gene sequences. Mol. Phylogenet. Evol. 43: 787-794.

Whitaker, R. and Basu, D. (1983). The gharial (*Gavialis gangeticus*): A review. J. Bombay Nat. Hist. Soc. 79: 531-548.

Annex 1: Trend of hatching success of gharial in the river system of CNP

Year	No. of Eggs Collection	No. of Hatchlings	% of Hatchling	No. of Hatchling Survival after 1 year age	% of Hatchling Survival after 1 year age
1977	592	438	73.99	NA	NA
1978	310	162	52.26	NA	NA
1979	543	294	54.14	NA	NA
1980	264	187	70.83	NA	NA
1981	259	64	24.71	NA	NA
1982	90	38	42.22	NA	NA
1983	296	124	41.89	NA	NA
1984	40	33	82.50	NA	NA
1985	158	116	73.42	NA	NA
1989	253	144	56.92	NA	NA
1990	395	237	60.00	NA	NA
1991	359	281	78.27	NA	NA
1992	490	230	46.94	NA	NA
1993	428	280	65.42	11	3.93
1994	437	144	32.95	10	6.94
1995	221	97	43.89	17	17.53
1996	577	276	47.83	17	6.16
1997	311	106	34.08	20	18.87
1998	302	19	6.29	2	10.53
1999	408	101	24.75	10	9.90
2000	244	141	57.79	30	21.28
2001	291	81	27.84	27	33.33
2002	466	229	49.14	32	13.97
2003	347	169	48.70	3	1.78
2004	521	298	57.20	157	52.68
2005	510	333	65.29	80	24.02
2006	382	262	68.59	95	36.26
2007	343	117	34.11	53	45.30
2008	369	133	36.04	32	24.06
2009	101	71	70.30	41	57.75
2010	508	355	69.88	133	37.46
2011	634	256	40.38	141	55.08
Total	11449	5816		911	

Annex 2: GPS locations of gharial survey, January 2013

IDENT	LAT	LONG	DATE	ALTITUDE	River Width	River Depth	Habitat
251	27.55782872	84.12037463	20-JAN-13 11:01:15AM	145.00			
252	27.55530065	84.12032576	20-JAN-13 11:16:21AM	139.00			
253	27.54867098	84.13845255	20-JAN-13 12:32:31PM	117.00	121	2.1	Sand
254	27.54977672	84.13912822	20-JAN-13 1:27:13PM	120.00	151	2.7	Sand
255	27.55142360	84.13870996	20-JAN-13 1:42:55PM	133.00	242	1.2	Sand
256	27.54762417	84.13601090	20-JAN-13 1:49:31PM	118.00	121	2.4	Sand
257	27.54752652	84.13590873	20-JAN-13 1:49:41PM	117.00	121	2.4	Sand
258	27.54704791	84.13288915	20-JAN-13 1:53:22PM	143.00	181	1.6	water
259	27.55491718	84.12208764	21-JAN-13 11:55:36AM	116.00	303	2.4	Rocky
260	27.55071860	84.13128560	21-JAN-13 12:15:49PM	114.00	151	2.1	Rocky
261	27.55213363	84.13993824	21-JAN-13 12:59:53PM	113.00	121	2.1	Sand
262	27.55923947	84.14762654	21-JAN-13 1:30:09PM	116.00	242	1.8	Clay
263	27.56112171	84.14992050	21-JAN-13 1:42:57PM	115.00	212	1.8	Sand
264	27.56070111	84.14194051	21-JAN-13 1:52:37PM	113.00	545	2.1	Sand/rock
265	27.55082002	84.13837460	21-JAN-13 2:02:46PM	111.00	121	2.4	Sand
266	27.54725017	84.13319835	22-JAN-13 12:02:13PM	118.00	106	1.8	Rocky
267	27.54811803	84.13612163	22-JAN-13 12:12:38PM	114.00	121	2.4	Rocky
268	27.54999692	84.13741764	22-JAN-13 12:20:12PM	117.00	151	1.8	Sand
269	27.55024075	84.13910609	22-JAN-13	123.00	181	1.5	Sand

			12:45:40PM				
270	27.55515589	84.14450898	22-JAN-13 1:06:57PM	121.00	75	1.6	Sand
271	27.56158146	84.15032267	22-JAN-13 1:41:28PM	119.00	303	1.5	Sand
272	27.56124225	84.14322244	22-JAN-13 1:50:18PM	112.00	151	1.8	Sand
273	27.55310157	84.13868406	22-JAN-13 1:58:50PM	114.00	136	1.8	Sand
274	27.56289952	84.12367048	22-JAN-13 2:54:46PM	117.00	90	2.7	Sand
275	27.55110190	84.10951956	23-JAN-13 12:10:18PM	112.00	545	2.1	Rocky
276	27.55389593	84.09265366	23-JAN-13 12:22:56PM	113.00	606	2.1	Rocky
277	27.55747542	84.08213429	23-JAN-13 12:32:34PM	116.00	606	3	Rocky
278	27.56482928	84.06452576	23-JAN-13 12:48:55PM	122.00	181	1.8	Rocky
279	27.55166014	84.04737194	23-JAN-13 1:06:12PM	112.00	151	1.8	Sand/rock
280	27.53832260	84.03344650	23-JAN-13 1:23:21PM	101.00	136	3	Sand/rock
281	27.53608983	84.02517993	23-JAN-13 1:31:29PM	100.00	106	4.5	Sand/rock
282	27.54222362	84.01018756	23-JAN-13 1:49:50PM	99.00	181	2.4	Sand
283	27.54557907	84.00443707	23-JAN-13 1:55:14PM	98.00	181	1.8	Sand
284	27.54752945	84.00132855	23-JAN-13 1:57:33PM	99.00	121	3.9	Sand
285	27.54636336	83.97532236	23-JAN-13 2:16:02PM	92.00	151	2.4	Rocky
286	27.54833085	83.95596134	23-JAN-13 2:39:49PM	97.00	545	3.3	Sand
287	27.54334822	83.93617652	23-JAN-13 2:58:41PM	92.00	151	3.6	Sand/rock
288	27.53540059	83.91938447	23-JAN-13 4:14:47PM	92.00	121	3.6	Sand
289	27.52634478	83.92097267	24-JAN-13 11:02:26AM	141.00	90	4.5	Sand

290	27.51412019	83.92773234	24-JAN-13 11:30:29AM	92.00	90	4.8	Rocky
291	27.51703391	83.94302077	24-JAN-13 11:46:08AM	92.00	151	3.9	Rocky
292	27.52558580	83.95968709	24-JAN-13 12:05:55PM	92.00	151	3.6	Sand
293	27.52465842	83.96134185	24-JAN-13 12:08:28PM	90.00	151	3.6	Sand
294	27.52358763	83.96157922	24-JAN-13 12:25:17PM	90.00	212	3.6	Sand
295	27.51052938	83.96845917	24-JAN-13 12:49:01PM	89.00	181	2.7	Sand
296	27.50398228	83.96231540	24-JAN-13 12:59:58PM	90.00	212	3.6	Rocky
297	27.49328672	83.93889134	24-JAN-13 1:40:06PM	92.00	181	3.6	Sand

I want morebooks!

Buy your books fast and straightforward online - at one of world's fastest growing online book stores! Environmentally sound due to Print-on-Demand technologies.

Buy your books online at
www.morebooks.shop

Kaufen Sie Ihre Bücher schnell und unkompliziert online – auf einer der am schnellsten wachsenden Buchhandelsplattformen weltweit! Dank Print-On-Demand umwelt- und ressourcenschonend produziert.

Bücher schneller online kaufen
www.morebooks.shop

KS OmniScriptum Publishing
Brivibas gatve 197
LV-1039 Riga, Latvia
Telefax: +371 686 204 55

info@omniscriptum.com
www.omniscriptum.com

www.ingramcontent.com/pod-product-compliance
Lightning Source LLC
Chambersburg PA
CBHW031543210526
45464CB00003B/1132